Wh Zero?

By Betsy Franco

Consulting Editor: Gail Saunders-Smith, Ph.D.
Consultants: Claudine Jellison and Patricia Williams,
Reading Recovery Teachers
Content Consultant: Johanna Kaufman,
Math Learning/Resource Coordinator of the Dalton School

What's zero?
Zero is a number!
It looks like this: 0
Zero means there are none.

Zero is the number that comes before the number 1 on the number line.

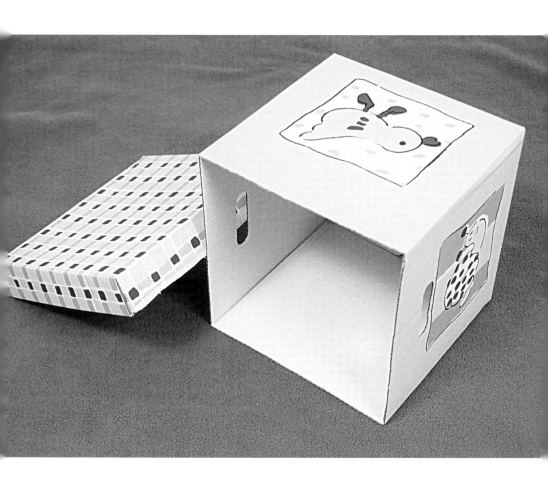

How many toys are in the box?
There are **0** toys in the box.

How many marbles do you see in this boy's hands?
You see **0** marbles in his hands.

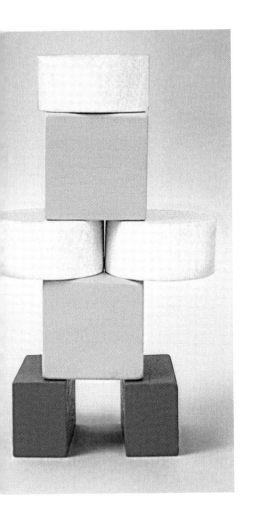

How many blue blocks do you see? 2

How many white blocks do you see? 3

How many yellow blocks do you see? 2

How many black blocks do you see? **0**

There are **0** black blocks!

In baseball, the score
"3 to 0" means one team
has 3 runs, and one team
has 0 runs.
Which score is higher?

Zero means there is nothing left.
These children ate
their cookies and want more.
There are **0** cookies
left on the plate.
These children have
no more cookies to eat.

Zero plus any number always makes that number. First there are **0** people in the park. The empty park is very quiet.

There were **0** children playing in the park. Then 5 children came to play in the park. How many children are now in the park?
0 + 5 = 5

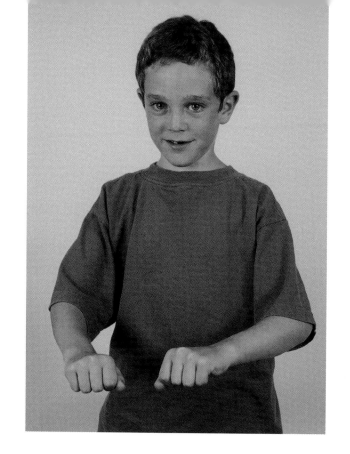

You can play games with **0**.
Which hand is holding 2 stones, and which hand is holding **0** stones?
Can you guess?

One hand has 2 stones.
The other hand has 0 stones.
How many stones are there in all?
2 + 0 = 2
There are 2 stones in all.

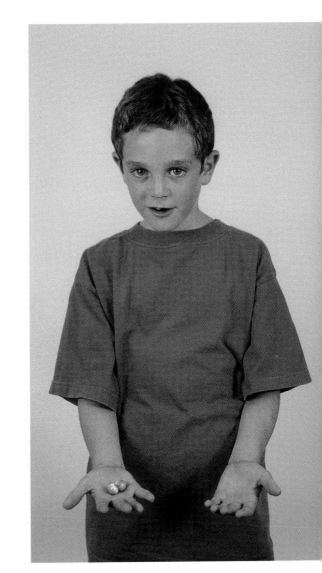

Take **0** away from a number, and you get the same number. Here are 5 happy friends, and **0** friends want to leave the fun. How many friends are left?

5 - **0** = 5

Five friends are left.

Take away a number from itself, and you always get **0**.
You have 25¢.
You buy 1 cup of lemonade for 25¢.
How much money do you have left?
25¢ - 25¢ = **0**¢
You have **0**¢ left.

All the bikes are lined up
in the bike rack.
Count them.
How many bikes do you see?
There are 4 bikes.

Four children ride their bikes home. How many bikes are left in the rack?
4 - 4 = **0**
There are **0** bikes left in the rack.

All the children have
gone to bed.
There are **0** children outside.

What do **0** children look like?
What do **0** children sound like?
Now you know what's **0**!